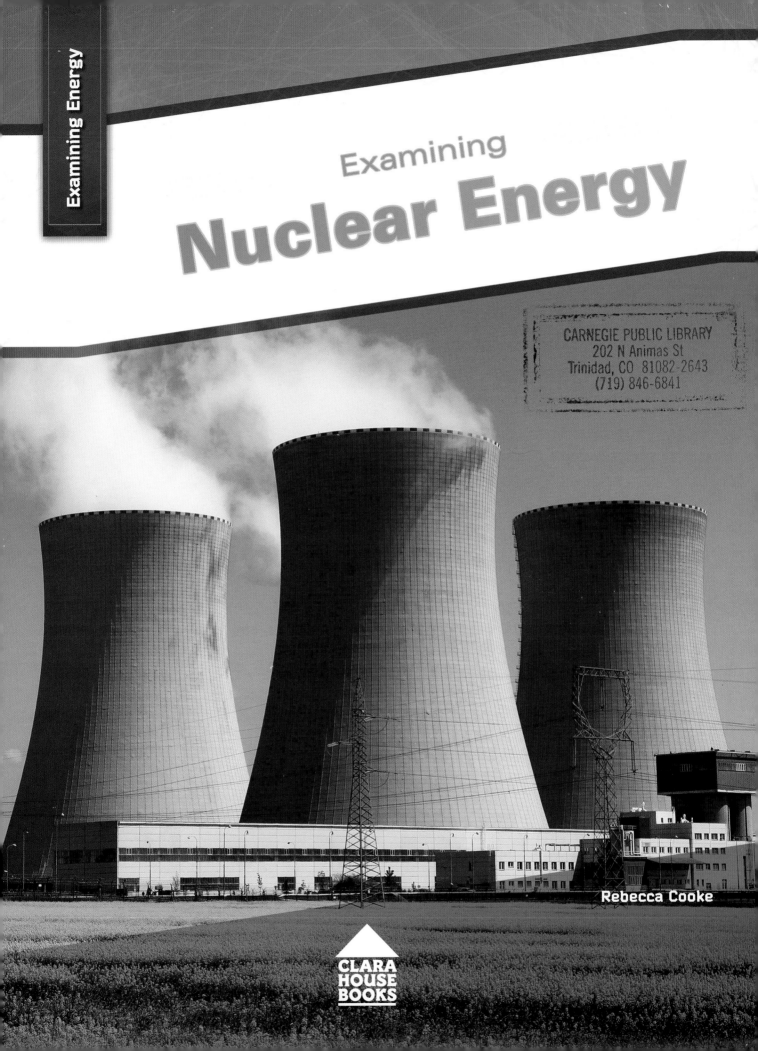

Examining

Nuclear Energy

Rebecca Cooke

CLARA
HOUSE
BOOKS

First published in 2013 by Clara House Books, an imprint of
The Oliver Press, Inc.

Copyright © 2013 CBM LLC

Clara House Books
5707 West 36th Street
Minneapolis, MN 55416
USA

Produced by Red Line Editorial

The publisher would like to thank Mikael Nilsson, Assistant Professor Department of Chemical Engineering and Materials Science, University of California Irvine, for serving as a content consultant for this book.

Picture Credits
Fotolia, cover, 1; Shutterstock Images, 5, 8, 14, 16, 27, 31, 34, 39; Andrea Danti/Shutterstock Images, 9; Sanden Sr/AP Images, 10; AP Images, 11; Anglian Art/Shutterstock Images, 13; Andre Luiz Mello/AP Images, 15 (left); Foreign Ministry, HO/AP Images, 15 (right); Olivier Tuffé/Fotolia, 17; John Carnemolla/Shutterstock Images, 19; Red Line Editorial, 20; Rick Gunn/Shutterstock Images, 23; Tomas Sereda/Shutterstock Images, 24–25; David E. Waid/Shutterstock Images, 29; Jan Zoetekouw/Shutterstock Images, 32; Daisuke Tomita/AP Images, 37; iStockphoto, 41; Library of Congress, 45

Library of Congress Cataloging-in-Publication Data
Cooke, Rebecca, 1978-
 Examining nuclear energy / Rebecca Cooke.
    pages cm. -- (Examining energy)
 Audience: Grade 7 to 8.
 Includes bibliographical references and index.
 ISBN 978-1-934545-43-0
 1. Nuclear energy--Juvenile literature. I. Title.

TK9148.C66 2013
333.792'4--dc23

                              2012036476

Printed in the United States of America
CGI012013

www.oliverpress.com

Contents

CHAPTER 1

A Nuclear Future?

It takes a lot of energy to power the world. Have you ever heard adults complain about the high cost of electricity? Right now, a lot of the energy we use comes from non-renewable sources. These non-renewable sources, such as oil and coal, can harm the environment, and they will eventually run out. Because of growing worldwide demand, energy sources of all kinds are becoming more expensive.

Scientists are constantly looking for ways to improve our energy sources. They want to find ways of producing energy that are more efficient, less expensive, and better for the environment than our current energy sources. Right now, more than 80 percent of our energy comes from fossil fuels, including oil, coal, and natural gas. These energy sources are made from organic material that has been buried underground for millions of years. In addition to being non-renewable, fossil fuels give off

Nuclear power provides one-fifth of U.S. electricity.

a lot of carbon dioxide and other types of pollution. Alternative energy research focuses on balancing our energy consumption needs with the needs of our environment.

Nuclear power already provides 13.5 percent of the world's electricity, and it may soon be even more common. In the United States, nuclear energy accounts for approximately 20 percent of the nation's electricity. Coal makes up a little more than 34 percent. Nuclear power was once considered the future of energy. It is very efficient—it produces a large amount of energy using only a small amount of fuel. Much of this energy is harnessed into electricity. Nuclear power plants give off much less carbon dioxide than fossil fuel plants. However, nuclear power plants also create dangerous radioactive waste that must be stored and protected. Though they are very rare, accidents occasionally cause this radiation to be released into the environment. Scientists are working to find safer ways to build nuclear plants and store waste, but these efforts require cutting-edge research.

EXPLORING NUCLEAR ENERGY

In this book, your job is to learn about nuclear energy and its role in our energy future. Is it safe? Should we increase our use of it? Can it be improved? Can it be effectively regulated worldwide? Terrence, Maria, and Lindsay are three teenage journalists reporting for the youth magazine *Stellar Science*. They are investigating nuclear energy and recording their findings. Reading Terrence's journal will help you in your research.

CHAPTER 2

The Amazing Atom

Alot goes into creating nuclear energy. Maria, Lindsay, and I are meeting chemistry professor Dr. Brian Andrews at the University of Chicago. We've asked him to fill us in on the basics of nuclear power. He greets us, and we sit down.

He holds up a small foam ball with smaller foam balls suspended around it. "This is a model of an atom," Dr. Andrews says. "Everything on Earth is made of atoms. Atoms have a nucleus in the center. The nucleus is made of positively charged protons and neutrons with no charge. Tiny, lightweight electrons move around the nucleus. These electrons hold a negative charge. Overall, the charges balance," he says. "The number of protons in an atom decides what element the atom is. Two protons make helium. Carbon has six protons. Uranium has 92 protons. Early research in the late 1890s showed energy rays coming from uranium and a few other elements. Scientists

An atom consists of a nucleus made of protons and neutrons. Electrons orbit the nucleus.

called this energy radiation. Elements that behaved in this way were called radioactive. The scientists also realized the radiation was harmful in certain ways."

"Do all elements give off radiation?" Maria asks.

"No, just some, " Dr. Andrews explains. "But an element can have different forms. These different forms are called isotopes of that element. The element's number of protons is the same, but the number of neutrons is different. Approximately 99 percent of the world's uranium is the U-238 isotope. U-238 has 92 protons and 146 neutrons. Most of the other uranium isotopes are U-235. The U-235 isotope has 92 protons and 143 neutrons. "

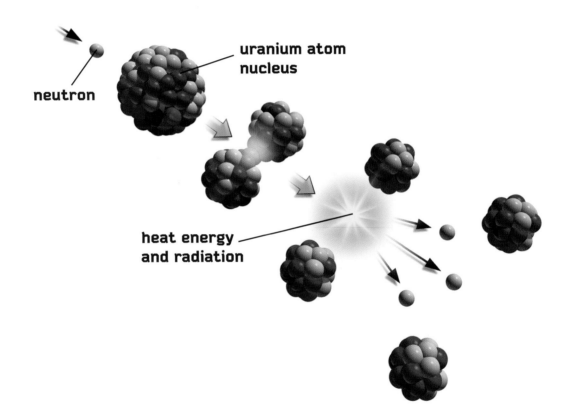

neutron

uranium atom nucleus

heat energy and radiation

NUCLEAR FISSION

Fission occurs when an atom splits into smaller parts. This split produces heat energy and radiation. Scientists have special techniques they use to make a neutron split. They can bombard a uranium atom with a neutron, which forces the atom to split. The particles thrown off hit other uranium nuclei. Then these nuclei split, creating a chain reaction.

Otto Hahn, *left*, and Lise Meitner, *right*, helped discover nuclear fission in 1938.

"Do the energy particles just shoot off from the uranium?" I ask.

"No, not like that," Dr. Andrews says. "Basically, we need to split up the uranium atom. When a uranium atom splits, particles are thrown off. These particles hit other atoms and cause those atoms to split. This process is known as a chain reaction. Imagine a line of dominoes standing up. When one domino falls, it causes the next one to fall. This is also a chain reaction."

Dr. Andrews tells us that splitting a uranium atom's nucleus releases a huge amount of energy. This is called nuclear fission. Fission happens naturally in a few elements, such as uranium and plutonium. During World War II (1939–1945), scientists used nuclear fission to create atomic bombs, two of which were dropped on Japan.

"While using the atomic bombs on Japan helped to end the war, people afterward tried to find peaceful ways to use nuclear energy," Dr. Andrews says. "They realized such massive amounts of energy could serve both constructive and destructive purposes."

"And that led to nuclear power plants," I said.

"Right!" Dr. Andrews says. I know where our next stop needs to be.

The atomic bomb blast at Nagasaki

THE BOMBING OF JAPAN

On August 6, 1945, the United States dropped an atomic bomb on Hiroshima, Japan. Approximately 70,000 people were killed instantly. On August 9, three days after bombing Hiroshima, the United States dropped a second bomb on Nagasaki, Japan. Approximately 40,000 people were instantly killed in Nagasaki. More people died from the effects of radiation as time passed. The day after the Nagasaki bombing, the Japanese began negotiating a surrender.

CHAPTER 3

Nuclear Energy in Action

We want to know how nuclear power is created. Today we are meeting nuclear engineer Janine Lindstrom in the lobby of the Comanche Peak power-generating facility in Glen Rose, Texas. Janine explains that the actual nuclear power plant is off limits, but she'll fill us in on how splitting a nucleus makes electricity. She'll also give us an idea of how a nuclear reactor works.

"Did you see the large, rounded dome on the way in?" Janine asks. "That was the containment vessel; it holds the nuclear fuel. The cooling tower lets off the huge, white cloud you saw as you were walking in."

We walk toward the large display on the left. It's a cutaway diagram of a nuclear reactor.

Janine tells us that Comanche Peak has two reactors. Inside each reactor, uranium fuel in the core splits apart into

Just one average-sized nuclear power plant can produce enough power to equal 1,200 windmills or 20 square miles (52 sq km) of solar panels.

smaller atoms, causing a chain reaction. As one atom nucleus splits, the particles thrown off hit other atoms, just as Dr. Andrews explained. Even though this process releases large amounts of heat, nothing burns.

I tell Janine about Dr. Andrews's dominoes example.

Janine smiles. "Very good! Here's another example. Imagine a box full of 100 mousetraps ready to spring. If you dropped one more set trap in, the motion of hitting the other traps would set that trap off. That motion would trigger the next trap to go off, and so on. That would be a chain reaction. If you compare this to nuclear fission, when the last trap is sprung, the fuel has run out. In atoms, the chain reaction in the fuel releases heat energy."

The chain reaction that occurs in nuclear fission is similar to the effect of a line of dominoes toppling over.

Janine tells us the power plant's core is made of uranium fuel pellets. For fuel, the plant uses U-235. This form of uranium will easily split and release heat energy.

The uranium is made into a black powdered uranium compound. Then it's formed into small ceramic pellets. These pellets hold of lot of energy. Just one pellet of uranium equals the energy from 150 gallons (570 L) of oil. These pellets are stacked inside 12-foot (4 m) metal rods, which are bundled together. The pellets and rods are called the fuel assembly. Reactors hold many fuel assemblies. All of the assemblies together are known as the core.

Janine points at a pipe in the model and tells us that Comanche Peak's reactors are pressurized water reactors (PWRs). The core of the reactor sits in water. The water flows through the core. The hot water flows to heat tubes full of water in a steam generator. The hot water in these tubes

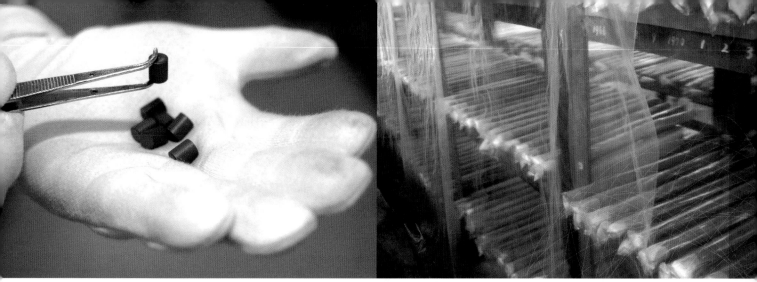

Uranium fuel pellets, *left*, fill fuel rods, *right*.

heats water to a boil in a separate container and turns it into steam. The steam powers a turbine, which runs generators, creating electricity. The water heating the generators is part of a closed system. Even though the water in the reactor's core is radioactive, it never mixes with the water making steam.

Janine continues, "A condenser catches all the steam, which moves to the cooling tower. There the steam is cooled until it changes back into water. Here is the best part: pumps then move the water back to be reheated and cooled. The water is used again."

IT'S NOT ALL ABOUT ELECTRICITY

The properties of radioactive materials have important uses besides generating electricity. Radioactivity is used to detect and destroy certain cancer cells. It can also damage normal cells, which means treatment must be planned carefully. Other forms of radioactive materials work as markers. If a patient has a medical condition a doctor wants to study, the doctor will sometimes have the patient injected with a weak radioactive marker. By following the path the marker takes, doctors can check organ or tissue function.

Janine tells us separate pipes cool the fuel parts in the core. "Electrical pumps circulate the water," she explains. "Backup generators run the systems in a power failure. In boiling water reactors, or BWRs, the same water is used over and over again.

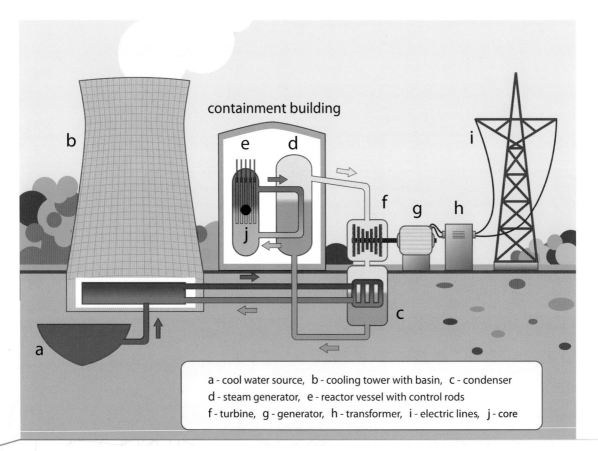

containment building

a - cool water source, b - cooling tower with basin, c - condenser
d - steam generator, e - reactor vessel with control rods
f - turbine, g - generator, h - transformer, i - electric lines, j - core

A NUCLEAR REACTOR

In a nuclear power station, the containment building holds the assembly with the uranium fuel. Control rods are inserted into the reactor vessel. Pumps move water or steam through pipes and out of the containment building. This steam turns a turbine to generate electricity, and the steam is condensed at the cooling tower and returned to its source. Then cool water is pumped back to the reactor, where the process is repeated.

It's heated by the core and changes to steam, which then powers the generator. The remaining steam goes to the condenser and then turns back to water to be reheated again."

Janine tells us the plant's workers keep the chain reaction under control by adding materials engineered to contain the heat and radiation to the reactor. Control rods absorb neutrons to maintain the heat generated by fission. They absorb energy, slowing the reaction. Water can also absorb the energy.

We thank Janine for showing us around. But my team still has a lot of questions. We wonder where uranium comes from and how it gets turned into fuel. We're on to our next stop!

A cooling tower removes the excess heat of a nuclear power plant and cools the steam back into water.

CHAPTER 4

Enriching Uranium

Radiation protection specialist Stephanie McCurdy greets us at the uranium enrichment plant in Paducah, Kentucky. Through the window we see a room resembling a factory with large containers and equipment. I think it looks more like a science lab than an enrichment plant.

"We enrich uranium here for use in reactors that generate electricity," Stephanie says. "99.3 percent of all uranium is U-238. Next common is U-235. It makes up only 0.7 percent of all uranium. U-235 is the kind of uranium used for fuel. The other isotopes are in very small amounts, but all uranium isotopes are radioactive."

"What does uranium look like?" Maria asks.

"The natural form of uranium occurs in compounds such as the mineral pitchblende. Pure uranium is a silver metal.

Open-pit mining is used when uranium is found near the earth's surface.

A baseball-sized chunk of it would weigh approximately 8.5 pounds, or 3.9 kilograms."

"Where does it come from?" I ask.

"Most of the uranium we use comes from ore fields in Australia, Canada, and Kazakhstan because those countries have the largest deposits," Stephanie says. She tells us uranium is often taken from open-cut mines dug into the ground. Deeper uranium deposits require underground mining,

FINDING URANIUM DEPOSITS

Early uranium prospectors easily located the element in ore bodies, or minerals holding ore, using radiation detectors near the surface called Geiger counters. Today, most high-grade deposits sit deep in the earth, buried in rock formations. Satellite imagery, geological and physical surveys, and complex analysis using computers locate possible ore sites. These ore sites are not only harder to find but they are also much harder to mine than the surface ore bodies.

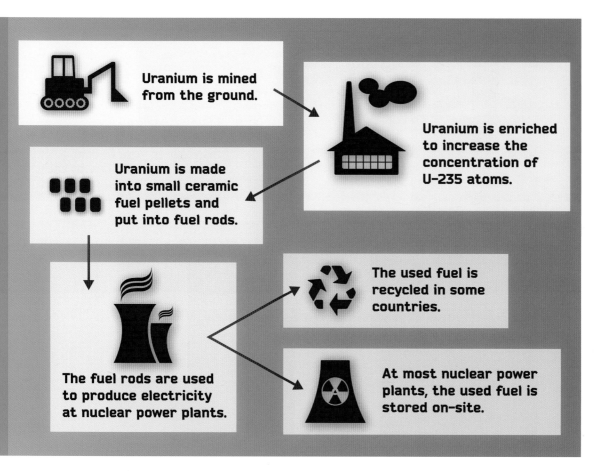

Uranium fuel cycle

using shafts and tunnels. Uranium is made up of different concentrations of isotopes, which is why it's enriched.

"How is it enriched?" Lindsay asks.

"We use a technique called gaseous diffusion," Stephanie says. "The processed uranium is called yellowcake. This yellowcake is converted to a compound. The compound is turned into a gas and filtered. The uranium compound gas is filtered until U-235 makes up between 3 and 5 percent of the total uranium."

Stephanie continues, "Other countries use a device called a centrifuge to spin the uranium very fast and separate the U-235 from the other uranium atoms. The United States is building more centrifuges for this process. The centrifugal technique is more efficient than gaseous diffusion; it requires less energy to enrich the uranium. Another new technique uses special lasers to separate uranium and pull out the U-235. No laser-separation enrichment plants are currently operating in the United States, but it is probably just a matter of time before we start using this technique commercially."

"Isn't it dangerous to handle enriched uranium?" I ask.

"It can be," Stephanie answers. "Workers wear special clothing and protective gear. Here at the plant, we face regular inspections to make sure we are following all the safety regulations."

We take a final look around the plant before we leave. I'm beginning to think each fact we learn brings up more questions. Still, it's interesting to investigate so many different aspects of nuclear energy.

CHAPTER 5

Getting Rid of the Waste

Today we're in geologist Maureen Larsen's office at the consulting group where she works in Reno, Nevada. We drove by the Yucca Mountain Repository area, but it was all closed up.

"I understand you're interested in Yucca Mountain and want to know more about the waste problems of nuclear energy," Maureen says. "Let me start by explaining a little about nuclear waste and storage."

She reminds us that all of uranium's isotopes give off radiation. This means they are constantly decaying, or breaking down. In time, a uranium isotope will give off enough radiation to become less radioactive. It will eventually decay into a different element. The decay is measured using the term *half-life*. A half-life measures the time it takes for an isotope to reach the halfway point in its decay to a different element. Uranium decays slowly. The half-life of U-238 is 4.5 billion

The Yucca Mountain Repository was proposed as a nuclear waste storage site, but President Barack Obama stopped the project in 2009.

years. This means it takes 4.5 billion years for half of any sample amount to break down to lighter elements. The half-life of U-235 is 700 million years. This is why U-235 is less common. Most of it has already decayed into lighter elements. Still, 700 million years is a long time. Nuclear waste is still radioactive. Right now, nuclear waste is stored on-site at each reactor.

"There's nuclear waste at every reactor?" I ask.

"Yes," Maureen answers. "Large cooling ponds hold the used fuel rods."

"Is it safe to keep the waste on-site?" I ask.

Maureen answers, "The pools are made of steel-lined concrete and hold huge amounts of water. The water shields the radioactivity. These ponds are located in protected places. After enough time, the waste can be moved to casks for dry storage."

Maureen tells us that the dangerous waste created by fission is one of the biggest drawbacks to nuclear power. Yucca Mountain was chosen for a national nuclear waste repository after the U.S. Department of Energy (DOE) investigated different ways to dispose of the waste. Geologists tested

Most nuclear waste is stored on-site at nuclear plants.

different materials to contain the waste and checked possible disposal sites. They considered placing sealed waste into the oceans. They looked at disposing the waste where the ocean and continental plates meet. The waste would have reentered the liquid mantle beneath the earth's crust. Other scientists looked at shipping the waste to the sun.

Maureen continues, "The chosen method was storage deep underground. Other ways proved too risky. Nevada was selected because studies indicated the waste could be stored here with less negative impact on the environment than other locations. Scientists began exploring the Yucca Mountain site by digging a tunnel. But problems came up. People understood the need for a place to put the waste, but many people in Nevada objected to having the waste stored so close to their homes. And to be stored in Nevada, the waste would have to be moved to Nevada. That affected people across the county who didn't want nuclear waste traveling by their homes. People feared that leaking, earthquakes, and other disasters might open the waste containers and cause radiation exposure. Eventually, legislation and lawsuits stopped the program. The DOE dropped the plans in 2009, leaving only the five-mile (8 km) exploration tunnel."

"Will waste storage continue at reactor sites?" I ask.

"For now—or until there's a central place for storage that everyone agrees on," Maureen answers.

RECYCLING NUCLEAR WASTE

In the United States, all used fuel rods are stored in storage ponds. Even though fission has stopped, the parts that held the fuel and its waste are still radioactive. The United Kingdom and France are two countries that currently recycle nuclear waste in reprocessing plants. After being reprocessed, the waste can be used as fuel again. But reprocessing plants are expensive to build, and many people wouldn't want a plant that processes dangerous radiation near where they live. Still, uranium is a non-renewable resource, so once it's gone, it's gone. Reprocessing can reuse as much as 97 percent of the spent fuel, helping our uranium supply last longer.

"Are there any new plans for nuclear power?" Maria asks.

"Probably. The last reactor was built in Tennessee in 1996, and newer reactors are in the works. Small reactors have possibilities, too. These have a core almost the size of a hot tub and are enclosed and buried. These mini reactors would supply energy for approximately 20,000 homes. Their size makes them portable. They could provide energy in locations where a large, expensive nuclear plant wouldn't be cost-efficient. They would need fuel replacements every seven to ten years."

We thank Maureen before heading out. Now that we've learned a little more about how dangerous radiation can be, we are curious about how nuclear plants operate safely.

La Hague nuclear fuel reprocessing plant in Normandy, France, recycles nuclear fuel.

CHAPTER 6

Regulating Nuclear Materials

To learn more about the safety of nuclear plants, we decide to meet with Vincent Georgio. Vincent is a Nuclear Regulatory Commission (NRC) safety inspection specialist in his Arlington, Texas, office. Vincent greets us, and we sit down.

"I hear you're interested in the safety of nuclear power," Vincent says. "Let me fill you in. Then you can ask questions as they come up." That sounds good to us!

"The NRC sets the standards, regulates, and monitors all nuclear materials," Vincent begins. "They are responsible for regulating anything related to nuclear energy, fuel, waste, and its transportation for commercial purposes."

He tells us the NRC was set up in 1974. The purpose was to make sure the public and environment were safe when radioactive materials were used. Today the NRC reviews the

Nuclear power can provide very efficient energy, but it also creates dangerous waste and can contaminate water.

safety and security of nuclear plants by handling all licensing, procedures, inspection, enforcement, and problem resolution.

Vincent adds, "We also regulate uranium mining, enrichment, and transportation. We monitor all waste and review licensing, including the Yucca Mountain waste repository. It is also our job to investigate new disposal techniques and monitor research. We work with almost every aspect of nuclear energy!"

Vincent tells us inspectors regularly check reactors, materials, and waste. They test the area around the nuclear

PREVENTING TERRORIST ATTACKS

After the terrorist attacks on the World Trade Center towers in New York City on September 11, 2001, the NRC ordered nuclear power plants to improve their security against possible future attacks. The NRC reviewed the security plans and the engineering of the reactors and containment buildings. Scientists studied possible events that might cause a radiation release, including airplane crashes. Experts experimented with the effects of various kinds of attacks and fires on reactors. They found that the reactors and spent fuel pools were designed to be safe enough to withstand most attacks. Still, each location developed a planned response that would provide immediate action if radiation leaked.

site for any problems or radioactivity. All workers, inspectors, and visitors to a nuclear site wear badges that monitor radiation. The inspectors also check the containment vessels, cooling pools, and waste at the reactors. Others inspect fuel assemblies and enrichment plants.

"Nuclear plants are designed to be as safe as possible," Vincent says. "The control rods slow the reaction. The reactor is inside a large steel pressure vessel with a water-cooling system. Everything is encased in concrete."

I remember hearing about how the nuclear reactors in Japan melted down after Japan's 2011 earthquake and tsunami. Could the NRC have prevented that? I ask Vincent.

"Japan was an unusual situation. Nuclear plants have backups in place, but in that case, the damage was so intense even the backups failed. All plants have a built-in shutdown. Backup cooling is automatic. Emergency generators provide

Engineers work hard to make sure nuclear power plants are operating as safely as possible.

energy to keep the water and pumps moving in case of a power failure. Computers and technicians monitor all these steps. Everything has a backup of some kind. We've worked with people in every field to use the best technology for our safety. The reactors are built to withstand extreme weather conditions. In the event of an accident, the reactors are shut down and people are evacuated. Security is very important. We are

Nuclear power plants are monitored closely to make sure everything is operating correctly.

working with dangerous elements. If a nuclear plant or nuclear waste fell into the wrong hands, it could mean trouble."

Our time has ended, and Vincent bids us goodbye. He's given us a lot to think about. The NRC takes many safety precautions, but if safety precautions fail as they did in Japan, that can be very dangerous. We decide to investigate further.

CHAPTER 7

A Nuclear Meltdown

Our last stop is the nuclear plant in Fukushima, Japan. After a 2011 earthquake and tsunami, this was the site of a major nuclear meltdown. We've been invited to join an international group of journalists on a tour led by Haru Sato, an officer with the Japan Atomic Energy Agency (JAEA). He'll summarize what happened to cause the meltdown and discuss the plant's recovery plan. Because the site still has dangerous levels of radiation, everyone on the tour wears protective suits and headgear. On the bus ride there, we listen to Mr. Sato explain what happened in two other nuclear disasters.

"The 1979 accident at the Three Mile Island nuclear plant in Pennsylvania didn't kill anyone," Haru says. "A malfunction caused the water supply to one of the reactor cores to shut off. A combination of equipment malfunction and human error

In 1979, the Three Mile Island nuclear power station in Pennsylvania suffered a partial meltdown. Only a small amount of radiation escaped, but the accident caused people in the United States to rethink nuclear power use.

meant the water was shut off much longer than it should have been. The power plant's reactor core overheated. The fuel pellets became so hot they actually melted the fuel rods holding them. This is known as a nuclear meltdown. This caused everyone to take a closer look at nuclear energy."

Haru continues, "In 1986, Chernobyl, in the Ukraine, had an even worse nuclear accident. Workers shut down the regulating systems in the core, and the chain reactions went out of control, leading to several explosions. The accident released radiation into the atmosphere. It killed 30 people initially, and more died from the radiation as time progressed. The city is now a ghost town."

Haru tells us most reactors in the United States and other countries have a different, safer design than the one at Chernobyl. Still the Chernobyl accident led to tightened regulations on nuclear power around the world.

"The Fukushima accident was different," Haru adds. "It wasn't caused by human and equipment error but by a natural disaster. During the tsunami, the backup generators flooded and couldn't work."

As we near Fukushima, we see deserted streets. Overgrown grass bursts from the sides of the pavement. As a twisted steel structure comes into sight, Haru tells us the Fukushima nuclear plant has been officially shut down. He explains the tsunami set off the chain of events that led to some of the reactors melting down and causing explosions. These reactors were boiling water reactors. After the earthquake, the safety systems worked at first. The plant's backup generators continued to cool down the reactors. The

DANGERS OF RADIATION

All radioactive material must be contained and handled with great care. Radioactive decay gives off three kinds of radiation, all of which can be harmful to humans. Alpha and beta radiation are particles resulting from radiation. These particles can move through some solid objects. Beta particles penetrate deeper than alpha particles. Gamma radiation is a wave. Radiation sickness results from exposure to radioactive materials, resulting in radiation burns and ulcers. Radiation damages bone marrow and can cause certain cancers. Radiation also damages genetic material, harming unborn children. Large doses can kill a person immediately, but most deaths occur from cancer over a period of time.

government evacuated people in a 1.9-mile (3-km) radius for safety reasons. Haru stops, pointing to the edge of the town we are entering. "But eventually generators flooded when the tsunami hit. Then they failed."

"What happened to them?" I ask.

"When the generators flooded," Haru says, "the batteries running the cooling lost power after some hours. The reactors began melting down. We had to plan to prevent three core meltdowns at once. Nobody knew Reactor 1 was already melting down. Reactors 2 and 3 would follow."

Haru explains that the tsunami created a state of emergency. Nobody could enter the site to make repairs. A few workers tried controlling the reactors by using seawater for cooling. But by then, some of the spent fuel pools were exposed. A fire and explosion caused further damage. Workers sampled the seawater for radiation and checked the electrical equipment as soon as they could return to the site. The plant workers restored some power and continued using seawater for cooling, but radiation levels remained high.

He continues, "We tried to control the reactors with various sealing and cooling efforts. However, the damage was already done. We discovered milk and spinach in the nearby area held radiation. The people evacuated had to leave their homes, maybe for years. They were exposed to radiation that might harm their health in the future. Crops are contaminated,

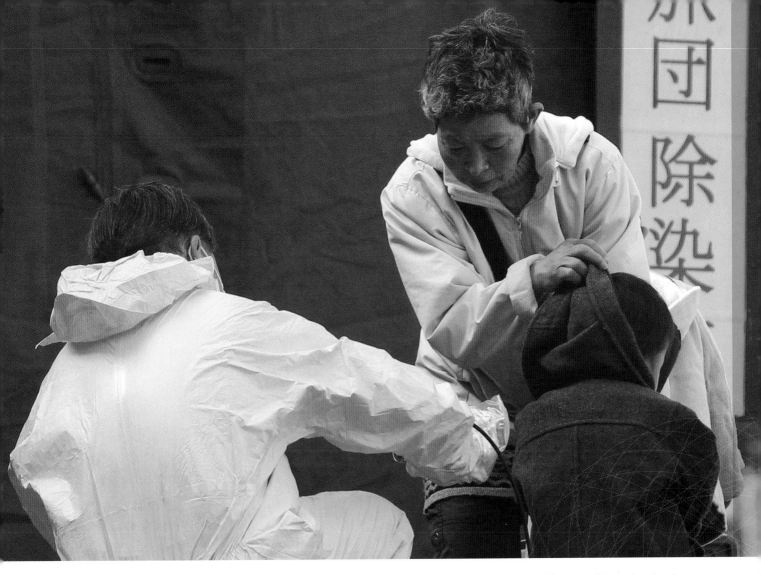

A worker checks a boy's radiation levels near Fukushima. The accident leaked dangerous amounts of radiation, affecting people, plants, and animals nearby.

and fishermen still do not know if the fish off the Fukushima coast are safe to eat. The loss of the power the plant provided will continue to affect businesses and industry due to reduced electricity."

"Is there a way to clean up all the radiation?" Lindsay asks.

"We must remove the spent fuel. That will take up to two years. We also must reduce the radiation levels in the surrounding area. Decontaminating the area will continue.

The reactors will be decommissioned, a process taking 30 to 40 years," Haru says.

He adds, "People have strong feelings against nuclear power here in Japan. Some countries, such as Germany and Austria, have made the decision not to use nuclear power. France plans to reduce its use. Other countries, such as South Korea and China, continue to build new plants. This is something each country must decide."

As the bus turns away from the plant, Lindsay, Maria, and I discuss what we've seen. We also review the booklet from the NRC. The United States believes their reactors are safe. After Japan's meltdown, U.S. nuclear plants immediately started working to improve their security. The NRC made plans for responding to natural disasters such as earthquakes or tsunamis. We read that nuclear energy continues to play a large part in the U.S. energy plan. Lessons learned from the three major nuclear disasters will continue to be part of the future as nuclear energy moves forward in the United States and around the world.

This trip has been an eye-opening look at the immense power of nuclear energy. I've learned that no energy source is perfect. And just like any other energy source, nuclear power has its pros and cons.

Despite its drawbacks, nuclear power is an efficient and important source of energy around the world.

Your Turn

You've had a chance to follow Terrence and his team as they conducted their research. Now it's time to think about what you learned. Nuclear energy comes from breaking the bonds of atomic nuclei. The energy released from splitting atoms heats water to create steam for producing electricity. Uranium fuel is found in nature, but it must be enriched to sustain a chain reaction. Nuclear energy provides a lot of power from relatively little fuel. But nuclear fuel creates a dangerous waste product. Many people work to keep nuclear energy safe, but accidents, while rare, still happen. These accidents can release dangerous levels of radiation into the environment, harming people, plants, and animals. As scientists keep working to improve nuclear power plants, their innovations will help determine how big of a role nuclear power will play in our energy future.

YOU DECIDE

1. Compare nuclear energy to one renewable energy source. Describe the differences. What advantages or disadvantages does nuclear have compared to the renewable source?

2. Nuclear power has many benefits as well as many risks. Do you think the benefits outweigh the risks? Why or why not?

3. What effects will stopping the production of nuclear power have on the countries that choose to use it no longer? Explain the effects of this choice and why you agree or disagree.

4. How big of a role do you think nuclear energy will play in the future of energy? Why?

5. What can you do to cut down on your energy use? Think about technology and ways to change your behavior, such as walking instead of driving.

How big of a role do you think nuclear power should play in our energy production?

GLOSSARY

atom: The smallest particle of an element that has the properties of that element.

carbon dioxide: A greenhouse gas released by the burning of fossil fuels.

chain reaction: A series of events where one event sets the next event in motion, such as a line of dominoes falling or nuclear fission.

core: The inner part of a nuclear reactor that holds the uranium fuel and control rods.

element: The smallest amount of a substance that cannot be separated into a simpler substance.

fission: The process of splitting an atom into parts.

isotope: Different forms of elements with masses that differ due to a different number of neutrons.

nuclear energy: The energy that comes from breaking apart the bonds holding atomic nuclei together.

nucleus: The center of an atom formed by protons and neutrons.

radiation: The energy given off during fission.

radioactive: The state of giving off energy particles.

reactor: The structure where atoms are split to produce heat energy.

renewable: When something can be replaced by natural environmental cycles.

turbine: A type of wheel that spins from a moving force, such as steam or running water.

uranium: A radioactive element that can be easily split to create nuclear energy.

The Uranium Cycle

Explain the path uranium follows to become the fuel for nuclear power. Follow each step from its point of origin to the enrichment process and then to a nuclear power plant and the eventual storage of the spent waste. Lay out the steps in a flow chart. Label each step. Then draw or find diagrams to illustrate each step along the way to show the complete process.

Shrink Your Carbon Footprint

The amount of carbon dioxide your actions produce is sometimes called your carbon footprint. Visit an online carbon footprint calculator to estimate how much carbon dioxide your household produces in a year. Examine your results. Where can you reduce emissions? Can you hang laundry in the sun instead of running the clothes drier? What about growing your own food to reduce driving trips to the grocery store? What are some other things you could do to reduce emissions?

Discover Nuclear Power's History

Investigate the beginning of nuclear power. Use this knowledge to learn how the early discoveries of radioactivity led to Einstein's theory of relativity and the use of nuclear energy in weapons. Start with the origins of nuclear power. Then study how the transition from its use in weapons to energy production came about.

Physicist Albert Einstein's theories were important in the creation of nuclear energy. Learning more about them can help you better understand nuclear power.

SELECTED BIBLIOGRAPHY

Alley, Richard B. *Earth The Operators' Manual*. New York: W.W. Norton & Company, 2011.

Cohen, Martin, and Andrew McKillop. *The Doomsday Machine*. New York: Macmillan, 2012.

Environmental Protection Agency. "Nuclear Energy." *EPA*, n.d. Web. Accessed May 10, 2012.

Mahaffey, James. *Atomic Awakening*. New York: Pegasus Books, 2009.

FURTHER INFORMATION

Books

Adams, Troon Harrison. *Nuclear Energy Power from the Atom.* St. Catharines, ON: Crabtree, 2010.

Metcalf, Tom, and Gena Metcalf, Editors. *Nuclear Power.* Detroit: Greenhaven Press, 2007.

Reynoldson, Fiona. *Understanding Nuclear Power.* New York: Gareth Stevens, 2011. 2009.

Websites

http://www.eia.gov/kids/
The U.S. Energy Information Administration site includes history, conversions, and general energy information.

http://www.nei.org/
The Nuclear Energy Institute's website offers information about nuclear power, including safety and waste disposal.

http://www.nrc.gov/reading-rm/basic-ref/students.html
The NRC site has a student section relating to current nuclear power.